W9-ABC-767

WE CAN READ about NATURE!™

A DRY PLACE

by CATHERINE NICHOLS

BENCHMARK BOOKS

MARSHALL CAVENDISH
NEW YORK

With thanks to
Peggy C. Hansen, Teacher,
Noxon Road Elementary School, New York,
for providing the activities in
the Fun with Phonics section and
to Beth Walker Gambro, Reading Consultant.

Benchmark Books
Marshall Cavendish
99 White Plains Road
Tarrytown, New York 10591-9001

Photo research by Candlepants Incorporated

Cover Photo: Corbis / Lanz Von Harsten, Gallo Images

The photographs in this book are used by permission and through the courtesy of;
Corbis : Chase Swift, 4-5; Lanz Von Horsten / Gallo Images, 6; Brian Vikanden, 7;
Peter Johnson, 9; Tiziana and Gianni Baldizzone, 10; Dave Bartruff, 11;
Julia Waterlow, 12; David Muench, 13, 14, 26; Steve Kaufman, 15, 25; Shai Ginotti, 17;
David A. Northcott, 18; Jonathan Blair, 19, 21; Anthony Cooper / Ecoscene, 20;
Joe McDonald, 22, 24, 27(top); Kevin Schafer, 23; William Dow, 27(bottom);
Bill Ross, 29.

Library of Congress Cataloging-in-Publication Data

Nichols, Catherine.
A dry place / by Catherine Nichols.
p. cm. — (We can read about nature!)
Includes index.
Summary: Introduces the people, plants, and animals that live in the desert and highlights
the ways that animals survive in this dry place.
ISBN 0-7614-1431-2
1. Desert ecology—Juvenile literature. [1. Desert ecology. 2. Ecology.] I. Title. II. Series.
QH541.5.D4 N535 2002
577.54—dc21
2001006240

Printed in Hong Kong

1 3 5 6 4 2

Look for us inside this book.

bat
bobcat
camel
fox
jack rabbit
kangaroo rat
lizards
quail
scorpion
tortoise

Deserts are dry places.

You can find deserts all over
the world.

Africa

Asia

There are even deserts in the Antarctic.

The ground is covered with ice, not sand.

But it is still a desert because very little rain falls.

Who lives in deserts?

People do.

Plants live there too.

Welwitschia plant

Century plant

13

Cactuses are desert plants that store water.

Ouch! Their spines are sharp.

The spines keep animals from eating
the plants.

Many animals live in deserts.

A camel can go for days
without eating.

Its hump is full of fat.

The camel lives off this fat
when there is no food.

The desert is cool in the morning.

That's when lizards hunt for food.

A tortoise also looks for something to eat.

The desert sun is hot in the afternoon. A bobcat rests in the shade.

Other animals go underground.

A fox goes into its den to stay cool.

What big ears these desert
animals have!

Long ears help keep them cool.

Jack rabbit

Kit fox

Water is hard to find in the desert.

That does not bother these animals.

Kangaroo rat

They get water from seeds
and plants.

Gamel's quail

The sun sets on the desert.

Many animals come out now that it is cool again.

Bat

Scorpion

They are not alone.

A desert is a great place to explore!

Fun with Phonics

(Answers on page 32)

1. WHICH ONE DOES NOT BELONG?
Read these words from the story. In each group, four words have the same vowel sound. Find the one that does not belong.

A. dry, very, find, spines, ice
B. place, rain, great, all, shade
C. seeds, these, eating, keep, help
D. afternoon, go, too, who, cool

2. WHO AM I?
Use the clues to name these desert animals. Write your ideas on a separate sheet of paper.

A.
Clue 1: I am a five-letter word.
Clue 2: My last letter is L.
Clue 3: I have two syllables.
Clue 4: I have two short vowels.
Clue 5: My third letter comes after L in the alphabet.
Clue 6: I can go for days without eating.

B.
Clue 1: I am a six-letter word.
Clue 2: I start with the letter L.
Clue 3: I have two syllables.
Clue 4: My first syllable has the vowel sound in "big."

Clue 5: My third letter is the last one in the alphabet.
Clue 6: Like the tortoise, I hunt for food in the cool mornings.

3. CHALLENGE: A DRY PLACE PUZZLE

Write the answer to each question on a separate piece of paper. The answers are all words in the story. Circle the last letter of each answer. Together they spell a secret word.

A. A fox might make its den here.
B. This turtle looks for food in the cool mornings.
C. These plants store water.
D. This covers the ground in the Antarctic.
E. There is very little of this, so animals get it from seeds and plants.
F. This animal rests in the shade. Its first syllable is a boy's name.

Fun Facts

- A desert gets less than ten inches of rain each year.
- The Sahara is the world's largest desert. The temperature there can reach 120 degrees.
- The Welwitschia (well-WITCH-ee-a) plant grows in the desert. It can live to be one thousand years old!
- Dromedary (DRAH-muh-der-ee) camels have one hump. Bactrian (BACK-tree-uhn) camels have two.
- Large ears help keep some desert animals cool. The blood vessels in their ears carry heat away from their bodies.

Glossary/Index

cactuses   desert plants covered with spines   14

camel   a large desert animal with one or two humps on its back   16

den   the home of a wild animal   21

desert   a place with very little rainfall   4

hump   a rounded lump on a camel's back   16

spines   the sharp points that grow on cactuses   15

store   to keep in a safe place   14

tortoise   a turtle that lives on land   19

About the Author

Catherine Nichols has written nonfiction for young readers for fifteen years. She works as an editor for a small publishing company. She has also taught high-school English. Ms. Nichols lives in Jersey City, New Jersey, with her husband, daughter, cat, and dog.

Answers to pages 30–31:

1. **Which One Does Not Belong?**   A. very   B. all   C. help   D. go
2. **Who Am I?**   A. camel   B. lizard
3. **Challenge: A Dry Place Puzzle**
A. underground   B. tortoise   C. cactuses   D. ice   E. water   F. bobcat
Secret word: desert